The Best Mother Ever

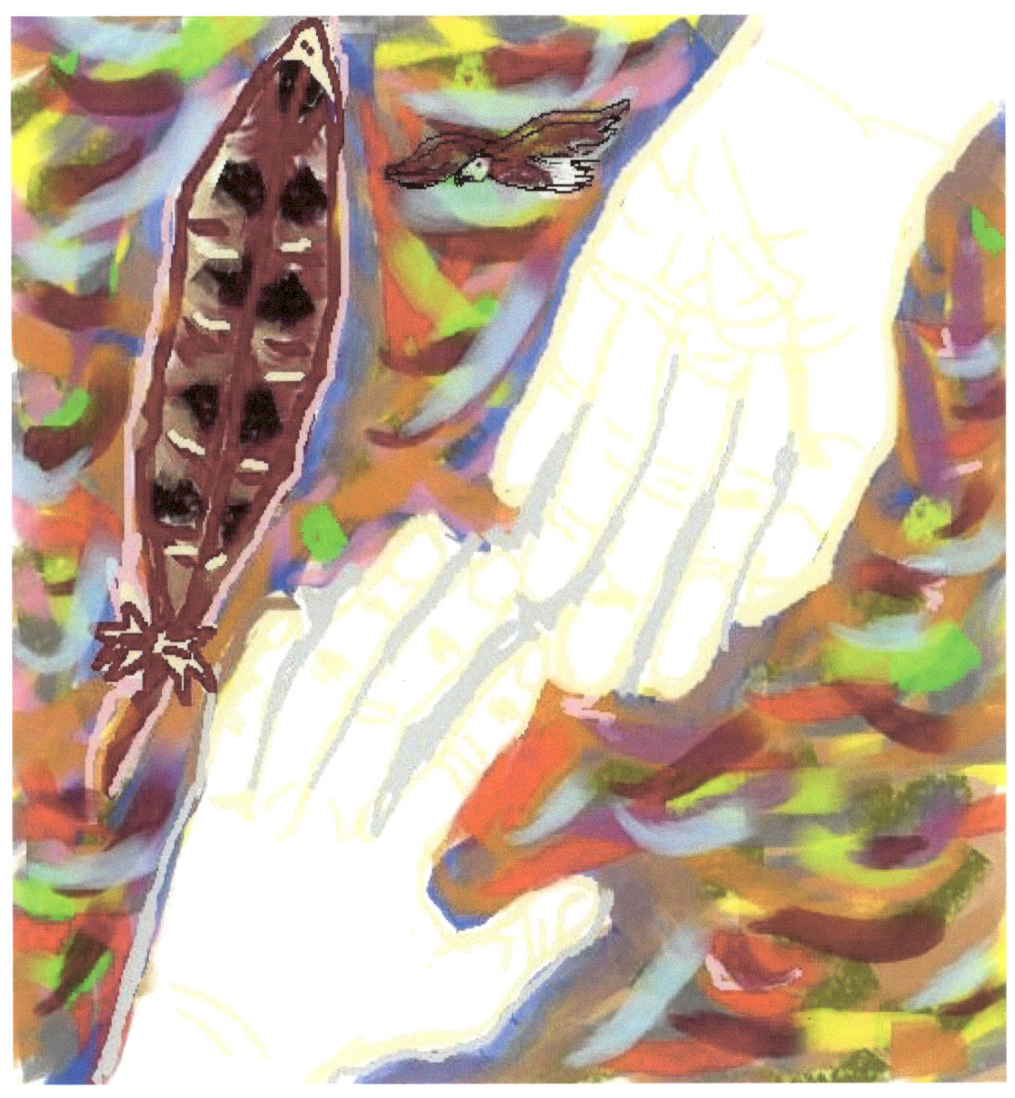

Written and Illustrated by G. F. Stone
©2011

The Best Mother Ever

Written and Illustrated by

G. Stone
G. Stone
© 2011

Printed by Spaulding Press, Bethel, VT.
All rights reserved.
Distributed by G. Stone
Website: Coming soon
email: gemmahstone@gmail.com; thebestmothereverbyg.stone@gmail.com

The Best Mother Ever

is dedicated to…
My children, Serah and Nathan,
My son-in-law, Tim,
My grandchildren, Parker and Taylor
And especially to Auguste.

Thank you for the experiences and guidance you provided the wisdom you shared and the dreams you helped come true.

My Feelings

    There are many people who are beginning to realize that some of their ancestors made some very big mistakes. Some treated the Indigenous people of the lands all across the nation very badly. They forced the Native people off of the land that they lived on, treated them as slaves, thought of them as barbaric and uncivilized, and took more and more from the Native people due to their own greed. The most important things that they could have learned from the Natives, they ignored, or misunderstood as foolish folklore.

The Indigenous cultures, or the ways that the Native American people of this nation lived before and still live, are very different from the way that many of us live today. In those cultures people get very close to the Earth as their Mother, and the Sun and Sky as their Father. The American Dream used to be what many Native Americans believe today. Everyone takes just what he or she needs and is satisfied. People take care of each other; families live together – grandmother, grandfather, mother, father, son and daughter, and they take care of the Earth they live on giving Her so much deep respect that She flourishes.

I have felt no greed from the Native People I know; prayer (especially prayer of appreciation) is a very important part of their lives. Silence and solitude are considered a central and essential part of Indigenous living. In this world we live in, because we do not pay attention to the Original People's culture, we are left with the need for attention, love, satisfaction, and approval. Most of all, we feel as if we need to be and are entitled to be taken care of, while we forget to take care of ourselves, our neighbors and our Mother–Mother Earth. For many of us, our attention is focused on how much we can get, and how fast we can get it. Many people today think that more is better, but they forget to stop, study and enjoy the important things in life...

The Plant People;

The Rock People;

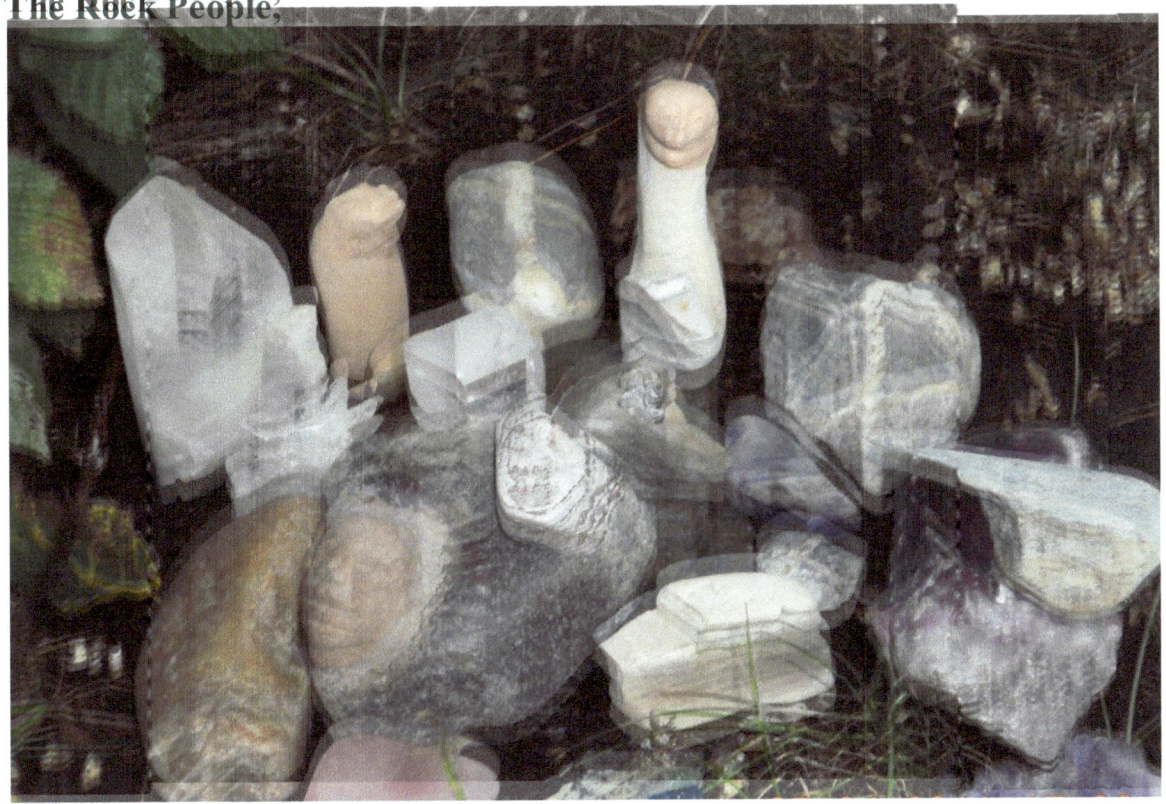

The Mountains;

The Ocean,

The Winged Ones,

Those who are Four-legged,

The Insect People,

The Creepers

And the Water People

These are the some of the very Special Beings that create true happiness and teach us what we need to know.

Today, there are some who have followed the original culture of the Early Americans their whole lives. I am fortunate enough to have a wonderful teacher who was able to teach me how to listen and hear the information that we should have heard hundreds of years ago, instead of listening to and learning from those who shared selfishness, self-righteousness and greed. It is because of this woman I am able to learn from the Best Mother Ever each and every day.

****

Prologue

One day, I was introduced to the best mother ever. She is a mom who is always there with me and for me. She is beautiful, wise, strong, gentle and loving. She is truthful and has a great sense of humor. She is the best teacher you could ever ask for, teaching lessons that will stay with me – and you - forever.

This wonderful mother has beautiful long hair like the long flowing strands of a willow tree. She has a soft lap to sit on, like the soft green moss you find on the forest floor or like the sand on the beach that conforms to your body no matter what position you sit or lie in. Her clothes have the richest natural colors you will ever find. They are green like the grass and trees, gold like the sun, blue like the sky and the ocean, orange, yellow and red like autumn leaves, pink like a lady slipper, deep maroon like a trillium, and brown like the soil and tree bark. Her clothes are so very, very pretty that they make me smile, and bring me deep peace.

   My mother is known as Mother Earth. I did not always know Her, but was introduced to her by a Wise Cherokee Woman. This is how it happened…

....

The Best Mother Ever

Chapter 1

The Little Girl and the Cherokee Woman

Once upon a time, there was a Cherokee Woman who knew that sometimes you just don't get everything that you need from your own mother; you see, that is impossible - mothers are human, and humans make mistakes. So, when I was just a little girl, I went on a long journey with this woman and she taught me the most valuable lesson I have ever learned. First, she asked me to tell her all of the things that I loved and were important to me. She sat quietly for a while as I thought, and then listened once I had the answer. I told her that the most important things to me are to feel loved, safe and cared for, and to be able to care for, protect and love others.

The Wise Woman said that I needed to meet someone. She said that she wanted me to meet the Best Mother that anyone could possibly have. I looked up at her and felt very confused because I knew that she was well aware that I already had a mother. She knew just by looking back at me that I did not understand what she had said. The Wise Woman said that Mother Earth could give me all of the things that I love and more, while at the same time giving me the opportunity to take care of and love her as well.

    The Cherokee Woman showed me how to sit quietly and listen so that I could actually hear Mother Earth talk to me. At first, I didn't hear a thing and I honestly thought that the Cherokee woman was being a little bit silly. But the Wise Woman sat very still, and remained very quiet to give me the chance to listen and actually hear what was going on around me. It was a very different experience than I had ever had before.

    Sometimes people are only aware of the things around them when they are frightened…but when they are aware of what is around them when they are scared, they are mostly trying to be aware of what they hope is not there. For example, there is no one behind me. No one is making any loud noises. There are no monsters under my bed. But listening while you're afraid is a very different way of listening and hearing than the kind the Wise Cherokee Woman was trying to teach and share.

    This Wonderful Wise Woman took the time needed to show me how to experience the tender loving care of Mother Earth until I was able to feel and experience it by myself.

She started by getting back to the discussion that we were having about the things that I loved. I said to the Cherokee Woman, "I want to see Mother Nature and feel Mother Nature's touch so that I always know that she is there." "There are many ways that Mother Earth touches us and lets us know that she is there," the Wise Woman said.

If you just look out the window, you will see her everywhere. There are trees, rocks, and plants - especially flowers. There is the sky, the clouds, the weather, the wind, the sun, ponds, lakes, and oceans …and those are just a few of the things that we can experience that have to do with Mother Earth every day.

The Wise Woman said that there are also ways that you can feel Mother Earth to know that she is there. She said that she holds and supports us like when we float in the water. She even creates soothing waves to rock us as we are floating. "Oh," I said. "I love when she rocks me in the water." Next the Cherokee Woman said, "Have you ever felt the gentle autumn breeze on your face?" "Of course," I responded. The Wise Woman said, "That is when Mother Earth blows you kisses and touches your cheek." "Oh…" I said.

Before the Wise Woman was able to get another word out, I quickly said, "But what about when I want to smell my mother's scent so that I can know that she is around." The Wise Cherokee Woman said, "Mother Earth has the best scent of any mother – in fact, she has many different scents like the different perfumes that your own mother wears." I really didn't understand what the Cherokee teacher was saying. She saw that I looked confused so she took me for a walk. We started walking up a mountain path in the woods, and the Cherokee Woman said,

"Close your eyes… now tell me what you smell." At first I said that I didn't smell anything, but the Wise Woman remained still and silent – she does that a lot. As I trusted her to lead me, I kept my eyes closed and began to become aware of the smell of pine trees. I quickly and excitedly opened my eyes, looked at the Wise Woman, gave her a big hug and said, "Mother Nature smells like the pine trees."

The Wise Woman smiled and told me to close my eyes again. As we walked, the ground was a little damp under our bare feet.

I not only smelled the lilacs that we had walked by, but I also began to smell the scent of damp leaves.

We walked a little further and she said, "I think we should take cover in this cave because it is going to rain." I didn't understand how she knew that it was going to rain, and she said that Mother Earth told her because there was a certain feel to, and a certain smell in the air…and the Wise Woman was right. Soon after entering the cave, it began to rain. The Cherokee Woman said that Mother Earth came to take care of us on this hot summer day because she knew we were feeling warm.

She said that she sent the rain to cool us and to water the little plant people.

"Mother Earth takes good care of us in so many ways," the Woman said. As the Wise Woman was talking I was watching what was happening after the summer rain.

First the sun peaked out from behind the clouds and then, the scent of Mother Nature, through the fine mist, was so refreshing and beautiful that I will never forget how She smells after a gentle summer rain – it was almost as if you could smell the sunshine, the mist, and each and every plant that was around us. I began to understand about the scent of Mother Nature.

I really enjoyed and was so very thankful for the lesson the Wise Woman taught me while I was walking up the mountain path with my eyes closed, but because there were so many things that I wanted to see, I opened my eyes to appreciate all that was around me.

Chapter 2
Mother Earth and the Milkweed Seed

I looked up at the Wise Woman and asked her if she knew how Mother Nature gives us comfort when we need it. And of course she knew. She has been listening and being close to our Mother her entire life. The Wise Woman asked me to tell her something that gave me comfort. There were two things that came to mind right away. I thought of how comfortable it felt to have a good friend. Someone who was always there when you needed her; someone who I could tell my innermost feelings to; and someone I could share all the good times that I had with. I also wanted someone to think enough of me to want to share those same things with me.

The Cherokee Woman smiled and told me that Mother Earth could be all of those things to me if I knew how to look and listen. She said that She had been a good friend to her, her entire life.

But, before I turned back around to face the Wise Woman, she was gone. I didn't know where she went, or how to get back to the trail. I was upset and began to panic because I thought that I was lost and all alone. But because the Wise Woman was my teacher, I thought about what she would have done and I remembered that the Wise Woman would have remained still and silent. So, I did the same.

I started to look at, and listen to the world around me and just think about how I felt. To my amazement, I began to relax a little. My friend, *Mother Earth*, was being still and silent just as the Wise Cherokee Woman would have been if she were there.

I then began quietly talking with nature. I started with the milkweed pod in front of me. As a tear fell from my cheek onto the ground, I told the milkweed pod that I was very upset that the Cherokee Woman had left me in the woods. I explained that I did not know how to get back to the trail so I could find my way back home and I was afraid. I told the plant that I wanted to know where the Woman was.

To my surprise, the milkweed pod split open just a little bit and one of her seeds squeezed out and caught a ride on the wind. First it floated back and forth like a feather or a leaf on an August day, and then it began to catch an updraft and slowly began to glide down the mountain.

   The seed blew to the left and then to the right. She then stopped for a rest on a tree trunk before she continued to move around one tree and then the next. I stood up and followed the milkweed seed and the wind as they seemed to be calling to me.

   As I walked I noticed the moss on the face of the trees. It looked so beautiful that I stopped and moved all the way around one great big sugar maple tree. I hadn't remembered seeing the moss on the way up the mountain and that was because the moss was mostly on just one side, the North side, of the trees. I realized that on my way up the trail, I had been walking in the opposite direction, so to get home I needed to walk down the hill and make sure that I kept the moss on the trees directly in front of me. I also began to realize that *Mother Earth* was always with me. She was there, teaching me how to pay attention to Her and learn the signs that show us which direction we are going.

Just as I was beginning to feel more relaxed and confident about finding my way back down the mountain, the Wise Cherokee Woman met me on the path. She had been watching me all along. She was still and quiet and allowed me to experience first-hand how Mother Earth was there with me to help me, as my good friend, when I thought that I was lost.

Chapter 3
The Comfort of Blankets

The Cherokee Woman then asked me what the second thing was that comforted me. I told her that the other thing that I thought of was my favorite blanket. When I'm tired or sick, the blanket feels warm, soft and cuddly. The Cherokee Woman understood and asked me if I could think of any natural blankets that could comfort me. I, too, was still and quiet as I thought for a while...

Then I said, "There is a blanket of silence at night, a blanket of darkness that is dotted with sparkling stars that are like tiny diamond-like night-lights suspended in the night sky. There is a blanket of fog, and a blanket of snow. There are blankets of moss on some forest floors that are very soft to lie down on, and blankets of flowers and pine needles that smell so very sweet ... All of them are comforting to me.

The Wise Woman said that those blankets are only some of the things that are comforting about Mother Earth. She said that in time, I would learn to feel, hear, taste, smell, see and sense all of them as I become more aware of my mother - Mother Earth. She said that I would even hear her sing me lullabies and songs when I hear the birds sing or the crickets chirp.

Chapter 4
The Mothering of Mother Earth

There were several other lessons that the Wise Cherokee Woman taught me about Mother Earth. She said that mothers take care of you when you are ill, and Mother Earth is no different. She grows healing herbs to take care of you – like chamomile[1], lemon balm[2], feverfew[3] and comfrey[4]. Have you ever gotten a mosquito bite and had it get terribly itchy? Well, Mother Earth grows little flowers called dandelions whose stems, when split open and then rubbed on the itchy spot help to stop the itching.

---

[1] Chamomile – An herb that calms your tummy when you have an upset stomach. It also relaxes you so you can sleep better.
[2] Lemon balm – An herb used in creams to help heal sores.
[3] Feverfew – An herb used to lower fevers, take down inflammation and get rid of headaches.
[4] Comfrey – An herb called bone-knit because it is said to help heal bruised and broken bones.

She also makes the sunshine warm you when you feel cold, and gives you water in the form of rain, rivers, lakes, ponds and oceans to cool you down when you are warm or have a fever. Mother Earth also salted the ocean to show us how salt water helps draw out some infections that we may have.

The Wise Woman also said that mothers tell us when to go to sleep and when to wake up. Mother Earth talks to the many Grandmothers and Grandfathers and asks that the Sun go down in the evening so we can go to sleep more easily. She asks it to rise and shine in the morning and has it wink at the birds to let them know that it is time to start chirping so we can gently wake. First one bird starts, then another, and another, and another until there is a loud cacophony[5] of sound. It is just like a wildlife symphony to help us wake-up.

---
[5] Cacophony – In this case it means that the birds and animals start singing so very loudly that it begins to sound like NOISE instead of sweet music. (and there would be no way that you could stay asleep)

Chapter 5
She Talks to Us

Throughout the years, the lessons that the Wise Woman taught were many. She said that mothers always know what to say, and again, Mother Earth is no different. She talks to us through the wind. She has the pine trees whisper to us to tell us what she wants us to know. She gives wind to the animals so that they can talk to us. She has the ocean wind talk to us as it causes the waves to crash and roll up to the shore and then to scurry back, like little plovers, as it rushes back to the Sea. She talks to us through Thunder and through the Birds and Insects who sing.

If we listen closely, she is talking to us all the time. It is so very important to hear a mother's voice, and I can hear Mother Earth's voice all of the time if I listen. I can even hear her voice in the silence...

Chapter 6
Home

As time passed, I learned that like all mothers, Mother Earth gives us a home. In fact she makes it possible for us to have many, many different kinds of homes. There are caves that some people use as homes, and there are trees to make stick or wooden homes. There is snow to make snow homes or igloos. There is grass to make straw homes. There are big and small rocks to make rock and stone homes. There is clay to make bricks so we can build brick homes. There are sides of hills to build homes into. … The land itself feels like my home.

Chapter 7
Protection

The Cherokee Woman taught me that Mothers protect us. She said that Mother Earth gives us Animals to protect us. She gives us "homes" to protect us. She teaches us lessons so that we can protect ourselves. Mother Earth gives us Mountains to protect us from high water or floods. She gives us Clouds and Trees to protect us from getting too much sun. Mother Earth truly cares. She <u>is</u> the Best Mother Ever.

Chapter 8
Smells

So far, the lessons that I have learned have been serious in "nature." But learning this next lesson about Mother Earth made me giggle a little bit:

Have you ever had people tell you that you were just like your mother? Sometimes people say that I have the same habits as my mother and that we look alike. Well, mothers have bodily functions just like all of us and… Mother Earth is no different. … Have you ever smelled a swamp or the ocean at low tide? It really smells stinky. It smells as if Mother Earth has passed gas.

(Tee-Hee!)

Just as there are lots of wonderful and sweet smells in Nature, there are also several "not so sweet" smells. Mother Earth is just like you and me. With the good smells there are bad smells and you wouldn't really appreciate the good smells if you didn't have the bad ones to compare them to!

****

Chapter 9
Mother Earth Teaches Us How to be Good People

The Cherokee Woman said that Mother Earth teaches us how to live with honor and dignity. She teaches us how to laugh at ourselves and how to be sensitive and compassionate with ourselves and with others. She teaches us about generosity and she teaches us how to work hard.

She said that Mother Earth sets a great example for us when it comes to having a good work ethic and having balance in our lives. Hmmm…let's see…There are sand storms to blow dirt around, and there are rainstorms to clean up the dust. When Mother Nature has worked hard making plants grow, there is a season for her to rest and a season for hibernation.

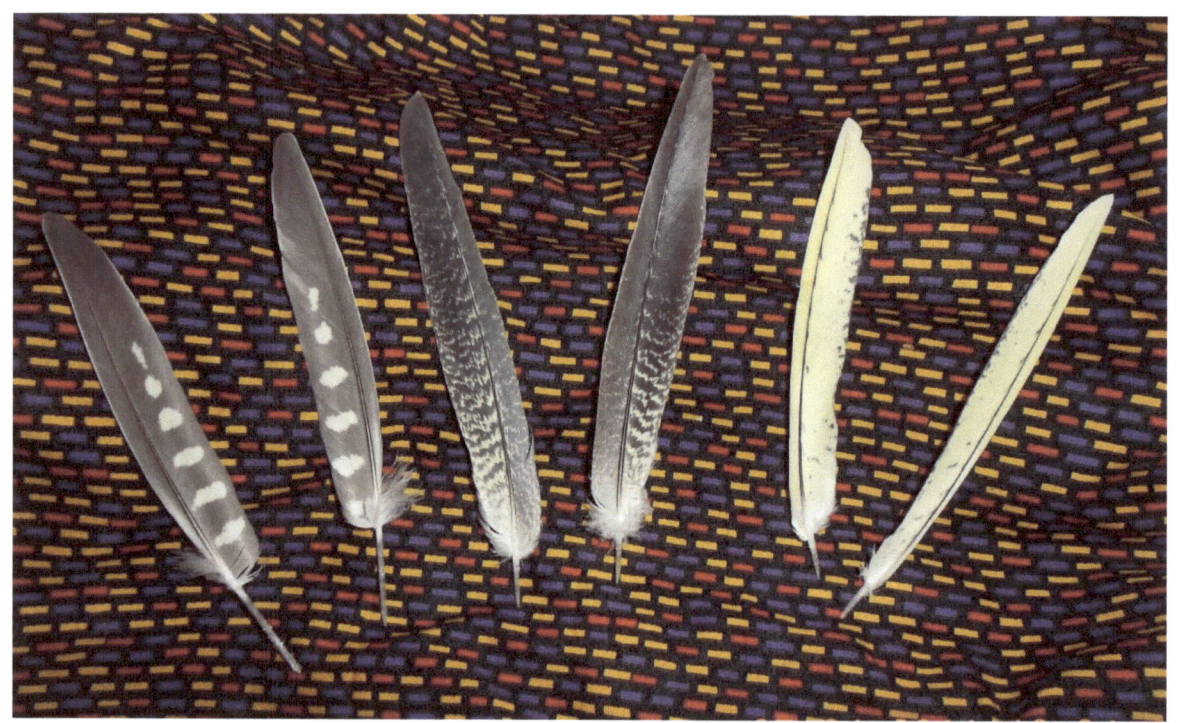

When birds preen their feathers, they will take out one feather on one side of their bodies, and then move to take out the feather in the same exact position on the other side of their body. If they pulled out feathers on only one side of their body, they might fly around in circles and not be able to fly straight. There are so, so many examples of this balance in Nature.

Chapter 10
Listening to Our Mother

Sometimes we have a hard time listening to our mothers. And listening to Mother Nature is no different. One day when I was out in my garden digging and working very hard putting in plants, Mother Nature thought that I had done enough and wanted me to go inside, rinse off and rest. I was not paying attention to what she was saying so she made it get cloudy. I still wasn't listening so she made it start to sprinkle and let me know that she would take the responsibility of watering the newly planted flowers and vegetables, but she wanted me in the house. Again, I chose not to listen so she made it rain harder. I still would not pay attention to her so it started raining so hard that the tiny water droplets began stinging my skin.

Owwwwwww!
Owwwwwww!

I started to get whiney and I just wanted to finish all the planting for the season ::: and this included planting all the herbs in the herb garden as well as planting all of the vegetables and flowers. By this time Mother Nature was so amazed that I was still not paying attention that she started to make it thunder and lightning as she tried to get my attention.

It seemed as if she was *scolding* me a little bit. Well, the thunder did get my attention and I finally realized what she was saying. "One should not hurry planting seeds in the garden." Planting requires care in what you are doing and I was tired and wasn't doing my best job. Mother Earth showed me that I should always try and do my best. In order to do this, I needed to go inside and get under cover, wash and rest and then continue the work - good work - on another day.

Chapter 11
Growth and Change

    As you can see from the story, over the years I began to listen to Mother Earth as my teacher. And as I listened more and more to Mother Earth, I noticed that the relationship that I had with the Wise Cherokee Woman changed from teacher/student to "Friend" and "Sister"…

  Instead of sitting with the Wise Cherokee Woman for lessons, we sat together more and more for tea and telling each other stories. Now, as I sit with her, in one solitary moment, I can see my friend and helper the milkweed seed blow by on her next journey; I can hear the messages of the wind in the pines and in the songs of the winged animals and insects; I can smell the comforting scents of Mother Earth and all while drinking of her healing and herbal teas with my Dear *Friend* and *Sister*, the Very Wise Cherokee Woman who taught me the greatest lesson of my life…*how to listen to and be with Mother Earth, <u>the Best Mother Ever</u>, every minute of every day.*

                        This is not the End

Epilogue

I have more stories to tell you about what I have learned from Mother Earth. I will do that in other books. I hope that you will enjoy the lessons and the stories just as much as I do. Mostly, I hope that you learn about the real American Dream, which is to take just what you need from Mother Earth and to remember to take very good care of Her. It is so important to do this so that we will have Mother Earth with us for a long, long time and so that your children and your children's children and all of their friends will get to know her as well.

About the Author

Gemmah Stone, although not Native American, is truly in love with the land and all of the Special Creatures that occupy it. She lives in Royalton, Vermont and as a retired teacher knows the importance of teaching children to love and care for the Earth and all of its inhabitants. This book is the first in a series of books that Gemmah has written about Nature. Subtle and not so subtle lessons are taught in all of the books that Gemmah has written. These lessons are ones that she has learned from Mother Earth because she was taught how to Listen:::

www.ingramcontent.com/pod-product-compliance
Lightning Source LLC
Chambersburg PA
CBHW051931210526
45473CB00006B/2210